Mauka to Makai

Hawaiian Quilts and the Ecology of the Islands

EDITED BY
Marenka Thompson-Odlum

Common Threads Press

Quilting the Ahupua'a

Mo'olelo: Stories from the Islands

Hawai'i: The Kingdom

Mauka: Mountains

Kula: Fields & Plains

Makai: The Ocean

Conclusion

Marenka Thompson-Odlum

Quilting the Ahupua'a

In 2020, I began working with the Hawaiian quilting group Poakalani & Company as part of a contemporary collecting project at Pitt Rivers Museum (PRM) funded by Art Fund. The aim of the project was to work collaboratively with artists, makers and practitioners to challenge the narratives often presented in museums about their specific cultures and/or localities. When I reached out to Cissy Serrao, current head of Poakalani & Company and daughter of its founders, I had a simple brief: the quilts should reflect the stories and knowledge that the group wanted to share with our audiences. The result was that the PRM is now home to fifteen new Hawaiian quilts.

The narrative of all the quilts together, both coincidentally and intentionally, depicts the concept of the ahupua'a. An ahupua'a is an ancient land division system used in Hawai'i, extending from the highest regions of the uplands (mauka) to the ocean (makai) and defined by varying geographic boundaries. The ahupua'a contains all the necessary environmental zones and natural resources to sustain life. Mauka and makai are terms still used casually in contemporary Hawai'i to denote relative direction: towards the mountains (mauka) and towards the ocean (makai).

The fifteen quilts capture various aspects of the nature, culture, history and ever-changing dynamic of Hawai'i. Some quilts traverse the vast Pacific Ocean, historically a pathway between the islands, with quilters finding inspiration in the marine life that nourished Hawai'i. Other quilts look to nourishment that comes from the land — in the form of fruits and ground provisions like kalo (taro) — while some pay homage to Hawaiian arts, such as feather work and hula. The narrative is both coincidental and intentional; Hawaiian cultural practices, knowledge and language cannot be divorced from the 'aina (land), and the ahupua'a being the primary land system is thus naturally and almost unconsciously expressed in all Hawaiian art forms.

Mauka to Makai: Hawaiian Quilts and the Ecology of the Islands hopes to take you on a journey through the ahupua'a via the quilting work of the Poakalani 'ohana (family). This book features the stories of the quilters and interviews with biologists, historians, farmers, activists and cultural practitioners whose work embraces a 21st century ahupua'a of sustainable stewardship practices. It looks at the coexistence of human civilisation with the earth, with minimal exploitation of resources to sustain the world's nations, peoples and economies. ▨

Marenka Thompson-Odlum is Research Curator (Critical Perspectives) at Pitt Rivers Museum, University of Oxford, where she is responsible for commissioning new objects for the museum's collections, building new relationships with Indigenous communities, and enhancing the museum's displays — a project that has led her to work with artists and makers from the islands of Hawai'i, Haida Gwaii and Hokkaido.

An Introduction to Hawaiian Quilting and the Poakalani ʻOhana

Cissy Serrao & Marenka Thompson-Odlum

Along with her sister Raelene Correia, Cissy Serrao teaches the traditional art of Hawaiian quilting at the Higashi Hongwanji Mission in Honolulu. Continuing the Hawaiian quilting class originally founded in 1972 by her parents, John and Althea Serrao, Cissy lives by her family's mission statement to preserve and appreciate the cultural legacy of Hawaiʻi's quilting tradition, and to teach it to anyone who wants to learn.

In conversation with Marenka Thompson-Odlum, Cissy shares her thoughts on the cultural significance and symbolism of quilting in Hawaiian culture, why it is so important to keep the patterns and tradition alive, and how she teaches this art form.

MARENKA THOMPSON-ODLUM: Tell us about Hawaiian quilting. How did it come to the Islands?

CISSY SERRAO: Hawaiian quilting is one of my passions. It started over forty years ago with my parents, who wanted to share this amazing art and tradition with anyone who wanted to learn. The history of Hawaiian quilting itself began when the first missionaries came over to Hawai'i. In 1820, the brig *Thaddeus* brought the first missionaries to the islands and within a few days, the first quilting circle was held on *Thaddeus,* with the royal women in attendance. The missionaries taught us how to appliqué and how to quilt, but their style of sewing and quilting as well as their designs had very little meaning to the Hawaiians. The missionaries had a patchwork style of hand sewing and the Hawaiians wouldn't adapt to that style of sewing; they didn't like cutting fabric into pieces only to sew them back together again. Eventually, the ingenious Hawaiians were able to find a new technique to create patterns that not only reflected Hawai'i, their island home, but also their traditions and culture. The Hawaiian quilt patterns are cut by folding the top design fabric to an eighth or quarter fold. It doesn't matter what colour fabric you use, the pattern was always cut out in one piece using the eighth or quarter fold that opened to a beautiful symmetrical design — similar to making paper snowflakes. The tradition included not only the fold, but also the design and the story it told. That is why my parents wanted to stay with that tradition and it is why we teach this style of quilting. Some people say that the Hawaiian quilt motifs that we see today on Hawaiian quilts may have come from the kapa (an old barkcloth fabric) and the designs that were printed on them. Between the 1820s–40s, patchwork-style quilts were more common. Around the 1860s, as new fabric arrived, the traditional Hawaiian quilt style that we see today began to appear across the Hawaiian Islands. Hawaiian quilts are so unique that you can go anywhere in the world today and you will recognise a Hawaiian quilt, and hopefully what it means and its tradition.

MTO: What does Hawaiian quilting mean to you, and why do you quilt?

CS: For me, Hawaiian quilting is a traditional art that was handed down to me through my parents from their parents, spanning over four generations. It's not just a generational tradition that I can now pass on, it's also part of Hawai'i's culture. The designs tell the story of Hawai'i's people, and some quilt patterns that my father designed even tell my story. The Hawaiians today are still trying to preserve, recapture and pass down traditions that have been lost. If I can add even a small part to preserving Hawai'i's culture and a tradition that comes from my ancestors, then we as a family and as a people will live on.

The number one thing about Hawaiian quilts that makes them so unique are the stories. All Hawaiian quilts tell a story. The stories are embedded in the pattern and even the fabric that is chosen. Who is it for? How long did it take you to make, six months or twelve years? What significant events took place when you made the quilt? It takes us a long time to make a Hawaiian quilt. You can't just make it in a day, or two days, or three days. It takes time. It frightens me that if you go out into Hawai'i, there are fewer people who teach the Hawaiian style of quilting as we do. It is a very small group. We don't want this tradition to stop and die; we want to keep teaching and passing it on. For us, by telling you these stories, it's not just about tradition or culture. These are actual beliefs in the Hawaiian system that the spirit is part of the quilt. When you make a quilt, your spirit becomes part of the quilt and with that we believe that it heals and soothes you. Every time a quilt is given to someone you love, you are giving love. You are giving love because it takes so much time to make a quilt. Another important factor of the quilt is the community of quilters. Hawaiian quilting guilds or classes bring people together. When you make a quilt, you put your mana (spirit) into the quilt, and I think that's part of the community, too, that energy.

"We don't want this tradition to stop and die; we want to keep teaching and passing it on."

MTO: What is the cultural significance of Hawaiian quilting today?

CS: The cultural significance of the Hawaiian quilt today is that it keeps some of Hawai'i's old traditions alive. Some quilts speak of old legends; there are some traditions on not only how to make a quilt, but even how and why we display them. Some quilt patterns are still inspired by the traditional use of meditation, prayer and spirituality. Here are some examples: when you finish your full-size quilt, you sleep with it for one night to seal your love into that quilt, because there are thousands and thousands of stitches in there and each stitch was sewn with love. A quilt made with love and given with love is love in its purest form. Another tradition is the echo quilting lines in quilts that encompass a border lei. For every quilted line that comes out from the centre of the quilt, there's a line that comes in from the lei border of the quilt. Eventually, these echo quilting lines meet — many people believe that it's the love coming out from the centre, from you, from Hawai'i. It then goes out into the world, and is returned. Another tradition we have is that you don't sit on the quilt, out of respect for the quilter. If you are in a house where there is a quilt on the bed, you lift the quilt and sit on the corner. Similarly, if you are having a party in the house, you display all the quilts on the beds so that your family members who own or made the quilts are there at the party with you. Every time we have a party, we say: "We've got to get out the quilts, so our family who have since gone will also be celebrating with us!" I don't think many people realise that the older quilts, the vintage and inherited quilts, tell the stories of old Hawaiian legends. For example, one quilt tells the story of how the Hawaiian people may have originated from India, and that's why we have the Indian breadfruit pattern called the 'ulu elekini — the Indian breadfruit has larger leaves and smaller fruit than our own 'ulu tree.

MTO: How do you feel about the fact that some of these historic designs are available through the internet and museums — that people are maybe copying them but they are doing so without thinking about Hawaiian tradition or respecting the kinds of things that you teach in your quilting school?

CS: I think that's one of the reasons why my parents started the quilting class. They wanted to teach others that Hawaiian quilting is not just sewing a quilt but that it also has its own traditions, restrictions and boundaries. They themselves would not have taken any historical patterns unless it was from their family's collection or if they knew the designer. They wanted to keep within the old tradition of quiltmaking but add their own family tradition as well. Those of us who were raised in Hawaiian culture know better — unless we share our knowledge with others about it, it is hard to tell them that they may be stealing. Just because the designer or family member is no longer with us, that doesn't mean the quilt patterns those people created are available for others to use.

I always think that those who work with museum collections are in a very fortunate situation. They can see the objects up close while also caring for them. I sometimes feel extremely emotional looking at a quilt. As a cultural artist, you connect to fellow artists, weavers and quilters on a very emotional level. Every stitch is so wonderfully made and so beautiful. Your imagination takes you to what they were thinking when they were creating the quilt. It's extremely important that we not only share a level of empathy, but that we can also collect these stories to bring them to life and share them with the general public, so that people respect the tradition.

MTO: Tell us about the quilts that have been made for Pitt Rivers Museum.

CS: The quilts took over two years to complete. One of the reasons for this, of course, was COVID-19, which completely shut down the fabric shops where we needed to purchase our supplies. Secondly, we were originally commissioned to create only one 90x90 inch quilt, complementing a feather 'ahu that was in the Pitt Rivers Museum. After several conversations with Marenka, it seemed that one quilt couldn't really tell the story of Hawai'i. So, we decided that we would make five smaller 45x45 inch quilts, but when word got out to our quilting classes about the commission, many of our quilters wanted to be part of this amazing project. That is why we decided to open the project to all the Poakalani quilters who wanted to participate. Therefore, the Pitt Rivers Museum will have fifteen new Hawaiian quilts in their collection. All fifteen quilts together tell an even broader history of Hawai'i. It reflects who we are as a people, our culture and traditions, but the quilts themselves also tell the personal stories of the amazing quilters who created them. While some of the quilts are being made in Hawai'i, others are being quilted in various regions of Japan. One quilt was even sewn on the ocean, in the region of Papahānaumokuākea.

This interview was first published in *Spaces of Care: Confronting Colonial Afterlives in European Ethnographic Museums* (edited by Wayne Modest and Claudia Augustat, 2023, Transcript Verlag).

Cissy Serrao is the founder of Poakalani & Company, a quilting guild and school in Honolulu, Hawai'i, originally founded in 1972 by Cissy's parents, Poakalani and John Serrao. Her parent's philiosophy in creating this guild was to preserve and appreciate the cultural heritage of Hawai'i.

Meet the

Poakalani Quilters

Lights of the 'Iolani Palace
Yuko Nishiwaki

I have been quilting for over twenty years with the Poakalani quilting class, first under the direction of Poakalani and John Serrao, and now Rae and Cissy Serrao. I'm so proud to be part of the class and have made more than fifteen quilts of various sizes.

When I was asked to make a quilt for the Pitt Rivers Museum, I was honoured. I was given a beautiful floral pattern, but I soon realised, due to my age, my fingers could no longer make the delicate quilting stitches. Though devastated, I still wanted to be part of this amazing journey, so I looked through my collection of quilts and found the one quilt that represented a place where I found friendship and joy — the 'Iolani Palace, where our quilting classes were held every Saturday.

The quilt that I donated to the museum is *Kukui 'o Hale Ali'i* ("Lights of the House of Royalty," referring to the 'Iolani Palace). The 'Iolani Palace was the first official residence to have indoor electricity, even before the White House. This quilt represents my love for the Palace and the friendship I found there at the classes where I came every Saturday.

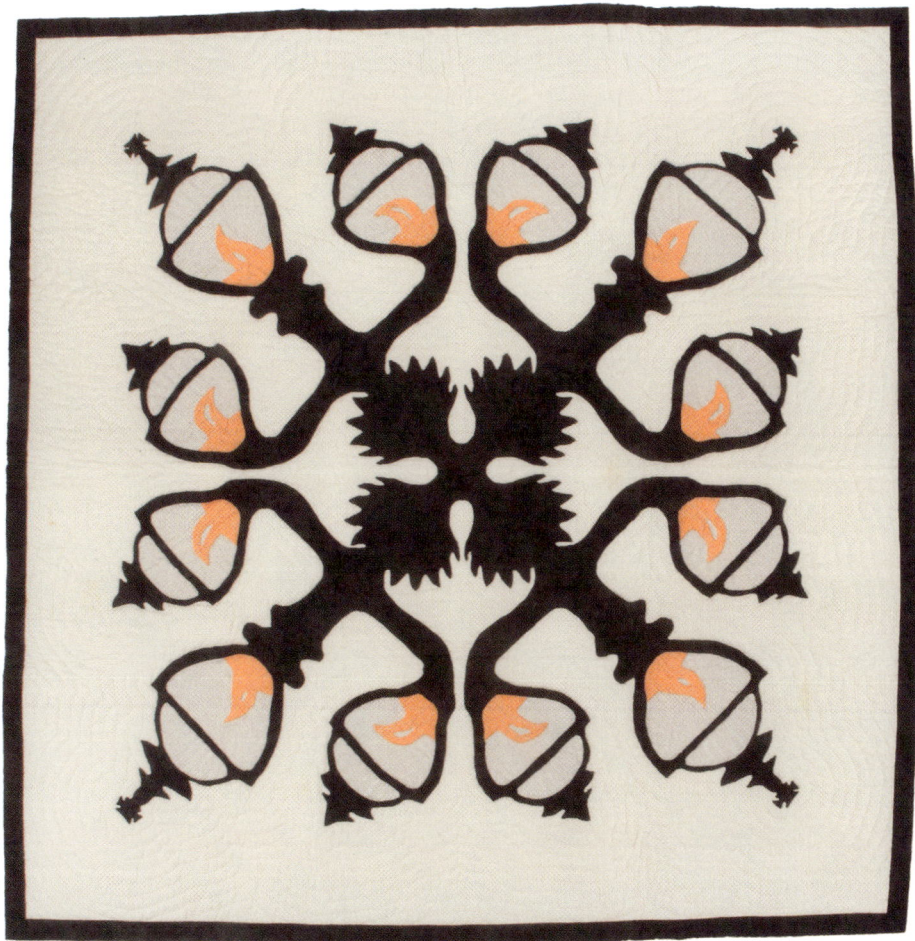

Nā Koa
Eriko Furukawa

It has been over twenty years since I first discovered Hawaiian quilts. The number of works I have produced is about thirty, ranging from large objects to cushions. I am originally from Japan.

My favourite quilt patterns are the "Hokule'a" tapestry and the "Lili'uokalani Thoughts" bedspread, both of which are wonderful patterns by John Serrao. He loved flowers and plants, but he preferred patterns relating to Hawaiian history.

I am very happy and honoured to be able to participate in this project and to be a member of Poakalani. Hawaiian quilts are an essential part of my daily life. The time during the day when the needle moves forward quietly is a special time when you can face yourself and prepare your mind. I have been able to be where I am today because of the support of many people around me, and I am still learning a lot from the Poakalani family — not only about quilting but also about the important aspects of being a person. I would like to continue this daily routine.

Nā Mea Ali'i Wahine
Nobuko Nakagawa

In 2000, I started living in Hawai'i with my husband. A friend took me to the Poakalani quilting class which was my first introduction to Hawaiian quilting. I learned how to make a quilt from Cissy Serrao and how to design from John Serrao. In 2001, I started designing quilts on my own, and in 2006, I obtained my instructor's licence and returned to Japan. I started teaching Hawaiian quilting classes in Japan with a total of four hundred students attending each year.

When I first laid eyes on the *Nā Mea Ali'i Wahine* design, the keyword "history" came to mind. That's why I wanted to try making a quilt with the same material that ancient Hawaiian women would have used. Polyester fibres were first introduced in 1953. Quilts made before then used cotton for both the batting and the thread. I experimented several times and ended up using cotton batting from Mexico and cotton thread from Canada. The fabric, however, is made in Japan. The quilt made entirely of cotton materials was slightly heavier and lacked elasticity compared to those made with polyester.

I chose the colours of the fabric based on the image of a kahili and 'ahu'ula displayed in Pitt Rivers Museum. I imagine that, illuminated by lamps in the dim museum, they would be truly mystical and beautiful. ▓

'Ōhi'a Lehua
Susie Sugi

I started quilting with the Poakalani family in 1998. I enjoyed making Hawaiian quilts designed by designer John Serrao because his designs always encompassed Hawaiian history, nature, sea life, and the Hawaiian spirit of 'ohana (family). I'm still making Hawaiian quilts for others, and special requests. I am a Hawaiian quilt lover and a member of the Japan Handicraft Instructors Association (Embroidery Section).

I finished the *'Ōhi'a Lehua* quilt using a two-combination method: Hawaiian quilt-making, along with embroidery for flowers and French knots for the pollen. I really enjoyed working with appliqué, embroidery, quilting and colour assorting for this design. For the lehua flowers, I used embroidery threads of fifteen different shades of red.

Having the opportunity to make a quilt for Pitt Rivers Museum is a great honour for me.

Kalo
Kimi Kumagai

I began Hawaiian quilting about thirty years ago. I have completed over seventy wall hangings and twenty bedspreads so far.

I currently live in Tsushima, Aichi Prefecture in Japan where I worked on the *Kalo* quilt for this project during the pandemic. Kalo (taro) is an important food for Hawaiian peoples; in Hawaiian mythology, it is said that kalo is the older brother of humans. An additional feature of this quilt is my use of the moa waewae (chicken feet) stitch to sew on the appliqué.

I am very grateful for being able to participate in this project. ▦

ʻUlu Poi
Tomoko Kato

I was born in Tokyo, Japan and my practice is based in both Tokyo and Hawaiʻi. I have been quilting for twenty-four years under the Poakalani and Serrao family, completing over one hundred quilts ranging from king-sized to smaller sizes.

When I saw Hawaiian quilts for the first time, I was inspired by the light shining from the heavens and seeing a path of light before my eyes, which became my starting point.

I run a PONO Hawaiian quilt class and have over one hundred students. While respecting traditional methods, we combine them with modern and unique designs, creating them in rich colours. Our works have been shared in magazines and on social networking sites. I would like to continue to design and create these exciting works.

ʻUlu is breadfruit. Hawaiian ancestors ate ʻulu, the fruit of the tree, as their staple food. It is a plant symbolising prosperity and fertility. ʻUlu is placed on a wooden board and crushed with a stone pounder to make poi. ʻUlu is still delicious to eat even today. In the centre of the quilt, I stitched the Kānaka Maoli, the native Hawaiian flag, to honour Hawaiian people and sovereignty. ▦

Nā Mele ʻo Hula Kahiko
Yoshimi Suzuki

I started quilting with Poakalani in 2000 at the Royal Hawaiian Shopping Centre. Under the direction of Kumu John Serrao, I was able to receive my teacher's certification in 2006. I started my own quilting class in Yokohama in 2006, called "Hālau Kuiki O Owyhee" (Quilt Class of Hawaiʻi), a name given to me by Kumu John. In the tradition of the Poakalani class in Hawaiʻi, I produced quilt shows in Japan every two years showcasing my students' quilts. In 2015, I received the highest honour from Kumu John: my Hawaiian name, "Pomaikalani," which means blessings from heaven.

The motifs on this quilt are the pahu hula (drum), ʻulīʻulī (feathered rattles), pūniu (coconut knee drum), all used in classical hula dancing, and maile leaves which represent Laka (the goddess of hula). In this way, they are related to Hawaiian life. The central part has wāwae moa (chicken feet) stitches, which is one of the traditional techniques.

When designers and quilters have pens or needles, we think about the lifestyle, history and traditions of Hawaiʻi. By reviving and expressing them, we play a role in conveying the history of Hawaiʻi, the aliʻi (royal family) and Hawaiian culture.

John and the Poakalani family have a history of learning the original techniques that were used on the ship *Thaddeus* boarded by missionaries in 1820. It's a great honour for me to learn from those who are the ancestors of this tradition, which is now loved all over the world.

We hope that people will be interested in the culture and traditions of Hawaiʻi through our quilts.

Ti Leaf and Laua'e
Pat Gorelangton

I have been a quilter for almost fourty years. Though I've tried other quilting styles, once I joined the Poakalani quilt group eighteen years ago, Hawaiian quilting became my passion. To me, it has a symmetry that appeals to my sense of order, while also having a beautiful freedom of movement, especially in the designs of John Serrao. John was a master designer; it was a privilege to learn from him. The knowledge and support that the entire Serrao 'ohana has given to me over the years is nourishment for my soul. One of my favourite things to do is to complete a quilt that was started by someone's grandmother or auntie years ago. Any time I make a quilt, whether as a commission, or just an idea I had and that I'm trying to express, it is truly a joy.

So many of the Hawaiian quilt patterns are inspired by our plants and culture. This *Ti Leaf and Laua'e* quilt is one of John's beautiful designs. There are two kinds of laua'e fern: one comes from Australia, and the other, much rarer one, is indigenous to the Hawaiian Islands. It has a faint but lovely fragrance and is used in gardens as well as floral arrangements. The ti leaf has many uses in Hawaiian life — from religious ceremonies and food wrapping to skirts for hula dancers and lei for ceremonies.

Pe'ahi
Anne-Marie Naughton

I joined the Poakalani quilting group more than twenty-five years ago. At that time, we were quilting at Queen Emma's Summer Palace in Nuʻuanu.

Poakalani herself was my first teacher and her husband John Serrao designed all my quilts. I was only planning to make a quilt for our queen bed, but after working with a beautiful red-and-white hibiscus pattern, I was hooked! I continued to make nine queen-sized quilts (90x90 inches), four twin-bed quilts (60x90 inches) and seven baby quilts (45x45 inches). I gave all my finished quilts to friends and family.

Poakalani and John's two daughters continued their parents' tradition of teaching Hawaiian quilting and helping us with our projects. This is much appreciated, as we love getting together to quilt.

I was born and raised in northern Sweden and later hired in Stockholm by Pan Am to be a stewardess. I was stationed in Los Angeles when my first flight took me to Honolulu, where I met my husband.

This quilt, *Pe'ahi*, depicts traditional Hawaiian crescent fans used predominantly by the aliʻi. These fans are made using intricately woven cords of coconut fibre, human hair, and plaited pandanus leaf. Intermingled with the pe'ahi design of the quilt are the fronds from the lauhala or pandanus tree, which is a key resource for Polynesian basketry work. Pitt Rivers Museum houses three pe'ahi, but elsewhere very few of these fans have survived intact.

'Awapuhi 'Ula 'Ula
Jennifer McCullough

I have been a hand quilter for the last ten years. In my work life, I am a bioacoustician (a researcher of animal sounds) so I spend many days at sea listening for whales and dolphins. While waiting for animals to produce sound, I hand quilt on the ship. In 2017, I moved to Hawai'i from San Diego, California and joined the Poakalani quilting group. John, Rey and Cissy Serrao taught me how to Hawaiian quilt and design my own patterns with a level of skill I never thought possible. My favourite quilt is one she and John designed called the *Flying Ocean*. It represents my love for the ocean, humpback whales, flying fish, flying squid and octopus.

For the exhibition, I worked on the *'Awapuhi 'Ula 'Ula* ("Red Ginger") quilt, designed by John Serrao, which is an iconic plant throughout the Hawaiian Islands. The pattern colour was chosen as a vibrant green due to the plant's beautiful leaves. This quilt was made during part of the COVID-19 lockdowns around the world. I took the fabric and pattern from Hawai'i to begin its voyage and spent two weeks in quarantine on the island of Guam in the Mariana Archipelago. From there, it boarded the NOAA research vessel Oscar Elton Sette for the next sixty days and travelled throughout the Mariana Archipelago — from as far north as the border waters of Japan, to transiting over the Mariana Trench in the south. Creating this quilt while at sea provided great calmness and comfort during a stressful time.

Bird of Paradise
Takako Jenkins

I have been a member of the Poakalani hawaiian quilting class for over twenty-five years. I've always loved crafts and sewing. When I moved from Ohio to Hawai'i, I decided to take a class on Hawaiian quilting and I've never looked back. The classes taught me not only about Hawaiian quilting, but Hawai'i's history and culture, and I met many new friends who I now call my family.

I have made over twenty quilts. John taught me how to design my own quilts, and my favourite is the one I designed for my husband. I showed John the pattern, but he told me I couldn't make the quilt at that time. He told me to wait and that he'd let me know when I could make the quilt. Ten years went by and I lost my husband due to illness and John finally gave me permission to make the quilt. Today, the quilt gives me great comfort.

I was so honoured to be asked to be part of this exhibit and happy to have quilted *Bird of Paradise* because of its regal colour and multi-coloured flower. My love for Hawai'i, my love for quilting, and the gratitude I feel to have met many close friends is reflected in this quilt. I love to quilt Hawaiian.

Honu
Hana Yoko Nakayama

I was looking for something to pass the time between dropping my granddaughter off at school at 7:00am and picking her up at the end of the day at 4:45pm. I took some classes at the Royal Hawaiian Shopping Centre, where I tried ribbon lei making and then hula classes, where I met Takako Jenkins. She told me that she joined a Hawaiian quilting class and that I should join too. So, I started my first class with the Poakalani 'ohana in 2003 and I'm still a member of that class.

I enjoy the classes because I love to sew and I met many amazing people in the class who are now part of my family. My favourite quilt that I made was the *Awapuhi 'Ula 'Ula* ("Red Ginger"). It is a memory quilt from when I visited Hana, Maui with the Poakalani family, where I received my Hawaiian name "Hana" from John Serrao.

I was so proud to be chosen to make a quilt for Pitt Rivers Museum. I was given the *Honu* ("Turtle") pattern. My granddaughter dyed the fabric, and on the shell of the honu, I quilted the flower for Hawai'i (the hibiscus), the flower for Great Britain (the rose) and the flower for Japan (the Cherokee rose).

Dolphins
Mie Tashiro

I have been part of the Poakalani Hawaiian quilting class for over twenty-five years, making over twenty quilts, and I have also received my teacher's certification. I do many types of Hawaiian crafts, including feather lei making, but I love Hawaiian quilting, especially the appliqué process.

Although I'm from Japan, I love Hawai'i, its history and its legends, but mostly I love the beaches and ocean. My favourite quilt is the one I made for my husband where John designed all the different fish that can be found in the Hawaiian waters.

I was so proud to be asked to make a quilt for Pitt Rivers Museum. I'm happy that the Poakalani family trusted me enough to be part of this commission. I was given a dolphin pattern, which I enjoyed quilting. My spirit and my love of the ocean and its dolphins lives in this quilt. ▦

Nā Iʻa o Ke Kai (Fish of the Ocean), clockwise from top left:

HEʻE by Susan Lessa
KALA by Rae Correia
HUMUHUMUNUKUNUKUĀPUAʻA by Midori Andrews
PYRAMID BUTTERFLYFISH by Cissy Serrao

Nā I'a o Ke Kai

He'e
Susan Lessa

I've been quilting for about twenty years. When dear friends left the islands for the mainland, I decided to make a Hawaiian quilt for them. Never mind that I had no idea how to go about making this special kind of quilt, my efforts were then cut short when my partly finished work was destroyed in a fire. Fortunately, my husband discovered the Poakalani quilters and encouraged me to attend my first Saturday morning class, and I've never looked back. I've been a member of this lovely 'ohana for nearly twenty years now. I value every minute and every friend I've met around a quilting table.

I was honoured to be asked to work on a quilt as part of the exhibit at Pitt Rivers Museum. My block depicting an octopus was one of four featuring different marine animals (*Nā I'a o Ke Kai*). Later, when my nephew announced that he and his wife were expecting their first son, I thought: here's my chance to quilt all four sections. My family loves the colours and the care that went into creating this heirloom.

At the class, we welcome newcomers who want to learn this special craft with open arms, much the same way I was welcomed many years ago.

Humuhumunukunukuāpuaʻa
Midori Andrews

I have been making Hawaiian quilts for fourteen years. I made one bedspread (90x90in) of John's design, two 45x45in, and several cushion sizes (22x22in). The one I'm making now is the size of a bedspread.

My favourite of John's designs are *Anthurium* and *Shell Ginger*. My most memorable moment with John was when he gave me a butterfly design the day after we decided to move from Hawaiʻi to the mainland. He somehow already knew that I was moving.

This time I quilted the Humuhumunukunukuāpuaʻa, the state fish of Hawaiʻi. I took care to make the stitches of the quilt smaller so that the design was easier to understand. I hope this quilt makes people smile. The Poakalani family is so kind and friendly to me and makes me feel like a sister. Seeing our quilters really motivates me to make quilts.

Thank you to the Poakalani family.

Pyramid Butterflyfish
Cissy Serrao

I grew up watching my mother quilt; she loved quilting the patterns my father John designed for her. She and my father also conducted workshops and demonstrations and she was amazing to watch. When she then decided to teach formal classes, it was inevitable that I would eventually learn how to quilt and assist with the classes. It was always our belief that the Hawaiian quilting tradition should be taught and passed on to anyone and everyone who wants to learn to quilt. I am so grateful and humbled that I can continue the classes my parents started fifty-two years ago. We are still meeting weekly and will continue the classes as long as we are able.

I wanted to be part of the Pitt Rivers Museum commission but time was not on my side, so I made a smaller quilt as part of *Nā Iʻa o Ke Kai* ("Fish of the Ocean"). I chose the Pyramid Butterflyfish, a popular reef fish. While it's a smaller contribution, I am so honoured that our class became part of this project. I hope that everyone who sees the quilts will appreciate Hawaiʻi's quilting tradition, and see the beauty and culture of our islands through the quilts.

Kala
Rae Correia

The tradition of Hawaiian quilting has been passed down several generations in my family, so it was a natural next step for me to continue the classes for another generation. I'm so proud of what my family has done, especially my parents John and Althea Serrao, who were part of the Hawaiian quilt revival during the Hawaiian Renaissance in the 1970s. My goal is to teach and share this textile art with the world, and that is why I'm so honoured that we were asked to make these quilts for Pitt Rivers Museum. It's such an honour for me, my parents, the generations that came before them and all the quilters who are sustaining this art to be shared with the world.

I quilted the Kala fish, one of the smaller quilts on the *Nā I'a o Ke Kai* quilt. The Kala fish skin was used to make the pūniu (coconut knee drum) that was used in the hula. This quilt brought back memories of when I once danced a hula using the knee drum.

Hua ʻAi (Fruits), clockwise from top left:
ʻŌHELO BERRY by Chikako Asano
ʻŌHIʻA ʻAI by Kathi Nakayama
LILIKOʻI by Kaelene Foo
MANAKŌ by Wakako Shionoya

Hua ʻAi

ʻŌhelo Berry
Chikako Asano

My sister Wakako [p. 52] got me interested in Hawaiian quilting when she asked me to accompany her to a beginner's quilting class in Japan. To my surprise I found out it was a hobby I really wanted to pursue. I eventually became part of the Poakalani quilting class, where I could not only learn to make smaller cushion quilts but also larger bed-sized quilts. The Poakalani class was always a positive place to be — their encouragement and compliments gave me the confidence and courage to complete the larger quilts. I was grateful to have John as my mentor.

I made the ʻōhelo berry part of the *Hua ʻAi* ("Fruits") sampler quilt. The ʻōhelo berry is a popular fruit in Hawaiʻi, and makes great jellies and jams.

I was so proud to be asked to be part of this quilt commission for Pitt Rivers Museum. I hope the quilts are loved and cherished, because I believe they were all made with love and given with love — the way we were taught.

'Ōhiʻa ʻAi
Kathi Nakayama

I started quilting in 2009 after I saw the beautiful quilts my mother Yoko Nakayama was making. My first quilt was a ti leaf wall hanging and cushion. I love making smaller quilts as wall hangings for my friends and family.

I am so proud to have been chosen to make a quilt for Pitt Rivers Museum. It's a smaller design, but part of a larger sampler quilt, *Hua ʻAi* ("Fruits"). The design I was given was the ʻōhiʻa ʻai (mountain apple) which is a sought-after fruit in the island. It was nice quilting alongside my mother while she worked on the *Dolphins* quilt [p. 41].

Manakō
Wakako Shionoya

I first got interested in Hawaiian quilting when I saw a Hawaiian quilt and fell in love with its design and flowing quilting line. I'm very shy and didn't want to take classes by myself, so I asked my sister Chikako [p. 50] if she would take classes with me. That was back in the year 2000.

My first quilt was the 'ulu design, which represented abundance in life's necessities such as food and shelter. It was a wonderful experience. Eventually, we met the Poakalani family of quilters, who were so welcoming. The classes not only taught me about Hawaiian quilting but also made me fall in love with Hawai'i and its people. I met many amazing quilters and I'm so proud to be part of this quilting family.

I made the manakō (mango) design on the *Hua 'Ai* quilt and was honoured to be asked to be part of this quilt and project.

Liliko'i
Kaelene Foo

I was born and raised in Kalihi, O'ahu and have been machine quilting since I was fourteen years old, and Hawaiian quilting for about ten years. I grew up with the Serrao family and we consider each other family. I have made several pillows and smaller quilts, and I am currently working on a 60x60 inch baby quilt and a 45x45 inch quilt. I also have larger quilt projects that are in the works.

I'm so happy to be part of this project for Pitt Rivers Museum where I know the quilt will be handled with care and in a museum where it will be cherished. I was given the liliko'i (passion fruit) pattern to quilt, as part of the *Hua 'Ai* series. It's an introduced plant which is grown commercially here in Hawai'i, and is used to make juice and all sorts of jams and jellies. Liliko'i is one of my favourite fruits it's such a fun pattern to work with.

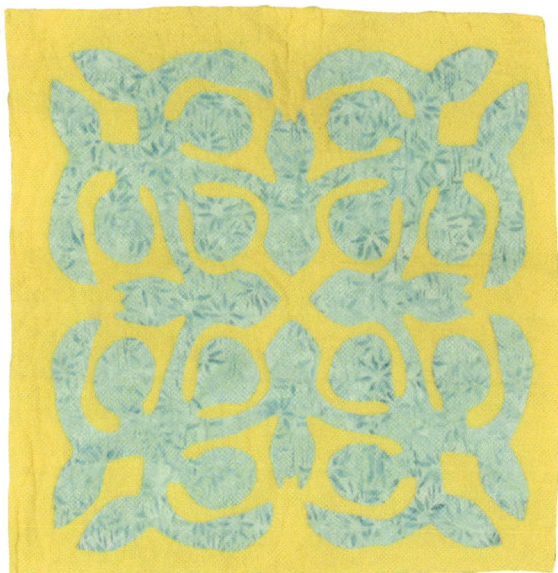

Hawai'i

The Kingdom

"The quilt that I donated to the museum is *Kukui 'o Hale Ali'i* ('Lights of the House of Royalty,' referring to the 'Iolani Palace). This quilt represents my love for the Palace and the friendship I found there at the classes where I came every Saturday."
— Yuko Nishiwaki

Kukui ʻo Hale Aliʻi translates to "Lights of the House of Royalty," and is referring to the lights outside ʻIolani Palace in downtown Honolulu. Today, Hawaiʻi is known as one of the fifty states of the United States of America. However, until its annexation in 1898, Hawaiʻi was a kingdom ruled by the aliʻi. The ʻIolani Palace is therefore the only royal palace on US soil. Construction on ʻIolani Palace began in 1879 and was completed in 1882. The Palace was the official residence of the Hawaiian monarchs, where they received dignitaries and luminaries from around the world who were entertained often and lavishly. The Palace was also the site of the imprisonment of Queen Liliʻuokalani, the last ruling Hawaiian monarch during the overthrow by the US government. After the overthrow of the Hawaiian monarchy, the Palace served as the capitol for almost eighty years, before it was later vacated and restored to its original grandeur in the 1970s. Today, the official vision of ʻIolani Palace is to be "a living restoration of a proud Hawaiian national identity and [...] the spiritual and physical multicultural epicentre of Hawaiʻi, representing the thriving dignity of the unique people of Hawaiʻi."[1]

Prior to the COVID-19 pandemic, the Palace was where the Poakalani quilters gathered every Saturday for class. As such, the Palace holds great significance to the quilters, and is a representation of Hawaiian history and sovereignty.

---

1   "Mission, Vision & Values," Iolani Palace, accessed February 25, 2024, https://www.iolanipalace.org/sacred-palace/mission-vision-values.

Ua Mau ke Ea o ka ʻĀina i ka Pono

A History of Hawaiian Sovereignty

Leilani Basham

This interview with Leilani Basham, Associate Professor of Hawaiian Studies at the University of Hawaiʻi, provides a brief look into the Kingdom of Hawaiʻi and the changes of the 19–20th century.

The history of Britain as a colonial power has involved the continents of America, Africa and many other places. Hawaiʻi is no exception. Captain Cook's[1] notions of the "discovery" of a people have long resulted in the complete discounting of indigenous spaces. Hawaiians have lived with that legacy; men like Cook are described in the history books as the "discoverers" of Hawaiʻi, even though they landed somewhere that already had a government and systems of environmental care in place. They were able to produce self-sufficiently, feed a million people, clothe them and house them. Cook himself observed that there was no one who was destitute, who was not cared for.

1   Captain James Cook, 18th-century British explorer and cartographer.

These prevailing narratives of discovery are not just that they first came to Hawai'i, but that they also brought their systems of hierarchy and ideas of superiority with them. What followed was the exploitation — economic, social and political — of peoples who were geared towards reciprocal relationships, who shared understandings of mutual giving rather than acquiring. It's how we ensure that people have everything that they need. If the other party believes in individual acquisition as their form of equality, through capitalism, then what we'll see is an uneven system. A system that would alter the fabric of Hawai'i.

Hawai'i had an established system of royalty: the ali'i (kings, queens, chiefs and chieftesses) and the maka'ainana (farmers and fishermen, literally "those that attend the land"). The ali'i had a responsibility to the people; if the people weren't fed, clothed or sufficiently resourced, it reflected poorly on the ali'i. They couldn't simply extort resources for their own desire or benefit. As such, a reciprocal relationship even existed between the different ranks in society.

This concept of reciprocity was also shared between the people and the land. Traditionally in Hawai'i, land was never bought nor sold. It was technically under the control of the sovereign leader, who would parcel out land to his ali'i. They didn't own it, but they had rights to be on the land and to the water. The responsibility of the ali'i was to make sure that those systems stayed in place. This means that everyone had access to land if they wanted to work, and that you have a responsibility and obligation to help care for things. For example, we have lots of fresh water in Hawai'i, and our kupuna (ancestors) created a waterway system where water could be shared. It went through man-made channels into the gardens, but then there were also pathways to ensure it went back into the stream — a system of recycling. These pathways meant there was a place where water is constantly flowing, and it's where we now grow kalo, our staple starch. The water picks up all of those nutrients from the kalo, and it goes right back into the stream. Maka'ainana had an obligation to maintain those waterways, but nobody was paid for their labour — it's just what you needed to grow food, to feed your family. We see ourselves genealogically connected to the land, water, taro. We are all one family. So, when foreigners arrived, we already had these well-developed agricultural systems. We had engineered these gardens and were able to successfully feed our communities.

In 1843, we temporarily lost our sovereignty to the British via Lord Paulet, who was on a diplomatic mission to Hawai'i. Under the guise of his so-called mission to protect British interests and citizens in Hawai'i, Lord Paulet unofficially and forcefully occupied Hawai'i for five months, wrestling power from King Kamehameha III. Paulet's act was later deemed unauthorised, however, and the British government restored Hawaiian sovereignty. At this reinstatement, the King first uttered the phrase, "Ua Mau ke Ea o ka 'Āina i ka Pono" ("the life of the land is perpetuated in righteousness") acknowledging that our sovereignty and balance had been restored.

Much of the British grievance in Hawai'i was over land, as the King had rejected several illegal land claims made by British settlers. Most of this stemmed from two very different ways of viewing land. Once foreign capitalism and the market economy was introduced, people wanted to know if they'd have land. Traditionally in Hawai'i, if you don't take care of your land, then your land will be taken away from you, but other than that, you're secure. The foreign use of land, however, was based on a constant need for more land. They wanted to be able to sell land, thus paving the way for private property. Our kūpuna were interested in these new technologies and new ways of being in the world, so we made a lot of adaptations. We privatised land in the 1840s–50s because we were told that's what "civilised" people do. "Civilised" people have real estate — it's no longer held communally. We then saw more and more foreign control of our land and economy.

Diplomatically, our ali'i were continuing to pursue means of restoring our independence. That was really their goal. This included establishing treaties with Great Britain and France in the early 1850s, thus establishing ourselves among this league of nations. If you have acknowledged that you have respect for my nation and I acknowledge that I have respect for yours, then that's a path by which we can maintain our sovereignty and our independence. And yet, notions of superiority were still at play.

By the late 19th century, there was a growing US influence. A lot of US citizens were moving to Hawai'i, trying to develop an economy that, for us, was already developed. Sugar cane and pineapple were the two main market economies that, at that point in time, were significantly growing markets. It came to a head in 1893 with the US presence in Hawai'i. Queen Lili'uokalani assumed the throne after the death of her brother in 1891. There had been a period where our King was forced at gunpoint to sign a new constitution, resulting in the loss of many monarchical powers that could control the government. Lili'uokalani started to receive petitions from the citizens to reinstate the old constitution, which would restore the powers to run her own government.

Foreign influence doesn't want that to happen, of course, and so they refuse to sign the new constitution, citing a fear for their safety. They requested that the US military came ashore to protect American lives and property, but they didn't locate themselves near the American contingent. They billeted themselves directly across from the 'Iolani Palace, directly in front of the government building. Just as the British had done in 1843, the Hawaiian monarch was effectively held at gunpoint. Queen Lili'uokalani followed the same example — abdicating the throne due to the threat posed by the US military — but she didn't entirely cede her power to those orchestrating the overthrow, American businessmen attempting to annexe Hawai'i. Instead, our kūpuna (ancestors) were very diplomatically strategic. They sent commissioners and delegations to the US Congress, making it clear that Hawaiians did not want this to happen. They had meetings with President Grover Cleveland, who denounced the US military presence as an act of war, but encountered a diplomatic stand-off, with the US ultimately choosing not to militarily reinstate the Queen. Two years later, in 1895, our people attempted to violently overthrow the government that came into power, but we were unsuccessful. My research looks at the songs of protest that were composed at this time. We had hundreds and hundreds of songs to protest these events — asserting our right to sovereignty and giving a voice to our resistance against US colonialism.

In 1898, the Spanish–American War broke out and there was additional pressure on the US to establish military outposts in the Pacific for the war effort. Rather than incorporating us through a treaty, then, we were annexed by the United States through a resolution process. This only required a simple majority vote by the US Congress, and gave no voice to Hawaiians. It's another sixty years before statehood is enacted, supposedly as part of the decolonising effort after World War II, during which time Great Britain was releasing many of its colonial possessions. In the case of the United States, however, rather than giving us a choice for independence, they incorporated us as a state, removing Hawai'i from the non-self-governing organisations list.

We've never acknowledged the right of the United States to have political power over us. They've even since apologised for their role in these events. Our resistance is something that we hold onto. We've never ceded our sovereignty.

Right: 'Iolani Palace, Honolulu, Hawai'i c. 1890
Original photograph courtesy of Library of Congress

Leilani Basham is Associate Professor at the Kamakakūokalani Centre for Hawaiian Studies, University of Hawai'i at Mānoa. Her research centres on understanding Hawaiian history, political and cultural knowledge, and practice from a Hawaiian perspective, using predominantly Hawaiian language resources.

Mauka

Mountains

"I chose the colours of the fabric based on an image of a kahili and 'ahu'ula displayed in the museum. I imagine that, illuminated by lamps in the dim museum, they would be truly mystical and beautiful."
— Nobuko Nakagawa

The quilt *Nā Mea Ali'i Wahine* depicts 'ahu'ula (feather cloaks) and kāhili (feather standards), which were both items used by the ali'i as an indication of one's rank. The quilt was inspired by the 'ahu'ula and kāhili housed in Pitt Rivers Museum (PRM). The 'ahu'ula were of particular importance, made from colourful feathers of native Hawaiian birds. They were worn in battle and used as diplomatic and political gifts. Although 'ahu'ula are primarily associated with male chiefs and royalty, the phrase "Nā Mea Ali'i Wahine" actually translates to queens or chieftesses (literally, "female rulers"). The quilt was thus named in recognition of Kekāuluohi, who gifted the cloak currently at the PRM to Sir George Simpson of the Hudson Bay Company in 1841 for services rendered to the Hawaiian Kingdom. Kekāuluohi was a powerful ali'i who served as Kuhina Nui (second-in-command) during the reign of King Kamehameha III and helped draft the first Hawaiian constitution.

Part of the allure of 'ahu'ula was that they were made out of feathers from native birds that lived in the mauka, or mountain forest. Because of their vibrancy and their elevated habitat, they were considered to be closely associated with the akua (gods) and thus the ali'i. Within the ahupua'a system, the uplands were a key habitat of native birds, and flora such as koa and olona were essential for making canoes and cordage. Many mauka flora and fauna are now under threat due to climate change, loss of habitat and mosquitos.

The following interviews dive into the importance of the 'ahu'ula in Hawaiian culture and history. They look at current threats to the survival of native forest birds, the steps being taken to save these species and innovative ways to tell their stories. 🔳

Stories Around a Feather Cloak

Shirley E. Buchanan
J. Susan Corley
Uluwehi Hopkins
Catherine ʻĪmaikalani Ulep

ULUWEHI HOPKINS: We're here today to talk about the feather cloak in Pitt Rivers Museum in Oxford that was given to Sir George Simpson, actually to his wife, Francis Simpson, by Kekāuluohi who was the Kuhina Nui of Hawaiʻi in the 1830s. Shirley, you've studied Kekāuluohi and her reign, so why don't you tell us a little bit about her and what Kuhina Nui means.

SHIRLEY E. BUCHANAN: Kekāuluohi is one of the most interesting women in history that I've studied. I came across her history when I was doing my own research in the Hawaiian archives. As it turns out, the actual state archive building is called the Kekāuluohi Building and it reflects a lot of what Kekāuluohi's life was about. She was Kuhina Nui from 1839–45 but she had been taught to be that from a much earlier age, so let's go back a step further to what Kuhina Nui means. The term roughly translates to the word "premier." In native Hawaiian culture, the Kuhina Nui position could be held by male or female aliʻi, who are considered the high-born royal family of Hawaiʻi.

What made Kekāuluohi so interesting was that she served at a really critical time in Hawaiian history. The Kuhina Nui had a power equal to the King, and if you were to go to the archives at the Kekāuluohi Building today, it mentions that this Kuhina Nui position was unique in the administration of the Hawaiian government. There was no equivalent in Western governments of the day; the Kuhina Nui held equal authority to the King in all matters of government, including the distribution of land, negotiating treaties and dispensing justice. The Kuhina Nui position pre-dated the 19th century. But what made Kekāuluohi's position so important was that this Kuhina Nui position was codified in the 1840 Constitution of Hawai'i.

UH: We should just note that the 1840 Constitution was the first written constitution in the Hawaiian Kingdom.

SEB: Yes. Throughout the first half of the 19th century, all of the Kuhina Nui positions were held by female ali'i (chieftesses). For me as a researcher, just seeing the names of women in the Hawaiian Constitution was an amazing thing.

So at the age of 15, Kekāuluohi was taught Hawaiian culture. She was learning how to negotiate with foreigners and how to observe the power structures of Hawai'i. She was also known to have an incredible memory. (We think she might have spoken multiple languages, but it's hard to know just from reading documents.) By the time that she gifts this 'ahu'ula, this feather cloak, she's been in a powerful position for quite a while.

UH: Catherine, why don't you tell us a little bit about women in Hawaiian society in general and then the difference between how women are perceived in Hawai'i versus, say, outside of Hawai'i at this particular time.

CATHERINE 'ĪMAIKALANI ULEP: Well, in Hawai'i, our mo'olelo — which roughly translates to "history and stories" — shows us our female Gods and our chieftesses were always known as strong and powerful. Hawaiian women within our mo'olelo always had the power to choose what they did with their bodies or who they chose to be with.

For example, we can see how Hiʻiaka took partners of both the same and opposite gender and that was completely acceptable, and the Makaʻāinana Wāhine were no different. Hawaiian women at all levels wielded the power to choose who they slept with, which was much outside of the norms of western Christian and patriarchal society.

SEB: This also makes Kekāuluohi really unique in that she was the wife of both Kamehameha I and, after he died, his son Liholiho or Kamehameha II. She didn't have children with either of them, but later became the wife of Charles Kanaʻina and had two sons. She was forty-one at the time which, in the

"At the age of 15, she was learning how to negotiate with foreigners and how to observe the power structures of Hawaiʻi."

context of the early 19th century, is an extraordinary thing.

UH: In all of these transactions, the woman had full autonomy; there are no forced marriages going on here. Catherine, you also studied the importance of clothing as political tools, gift exchanges or items of commerce. Can you tell us a little bit more about that? Specifically, what a gift of an ʻahuʻula might have been about?

CIU: Clothing is very different for makaʻāinana (everyday people), so we have to make that distinction even with ʻahuʻula or feather capes. Liholiho gave ʻahuʻula to people of similar status, such as the King of Britain. Chiefs also received items from King George and would often wear them to greet other sailors. We know that King Kamehameha received a sword because it's often written about. A Russian sailor, Otto von Kotzebue, wrote about how Kamehameha went on his ship and would dress in different types of western-style clothing. Many sailors from the early 1800s have also written about this; they talked about how Kamehameha would say, "well, I got this from 'my brother,' the King of England." And he wasn't the only one — ali'i wahina (chieftesses) did this too. They would go on the ships dressed very purposefully in nice, imported, western-style clothing. For the ali'i, this was very ceremonious. They were using it to display power and wealth. They wanted to show these people that they could not only dress nicer than they could, but also that they could harness power and wealth well beyond Hawaiian shores. So, they were using clothing for specific ceremonial reasons and they would go back and change into a loin cloth. When they were called on unceremoniously in their houses, these sailors were shocked!

SEB: I just want to add onto what Catherine is saying, because she mentioned this is well-documented in journals. There is one journal by Charles Wilkes in the 1840s where he reflects back on his trip. He says that the women came onto the ships in these big, beautiful hoop skirts and they had their uniformed beaus next to them. When we look at the histories of Kamehameha, we learn that when he was engaging with foreigners, he brought his wives onto the ships, so we can deduce that they are listening to all of the political and economic negotiations and discussions. By the 1840s, it is quite frequent that European men documented in their journals the detail of the clothing. Charles Wilkes, in fact, gives us a description of Kekāuluohi: her dress, the colour, the sleeves, all the way down to the kind of shoes that she's wearing.

UH: The fact that they made sure that appearances were illustrated is really interesting. Susan, can you tell us a little bit about the political situation at the time? What is the depth of authority that Kekāuluohi held at this particular time? Perhaps you can also speculate on why she gave this particular 'ahu'ula not to Sir George Simpson but to his wife.

J. SUSAN CORLEY: When Kekāuluohi presented the cloak to Sir George Simpson and told him it was for his wife, what she signalled was that it was a personal gift and that she wanted him to take it to his own home — from her home to his home. It was not a state gift, it was a personal gift.

Why would she make such a gesture? Well, for that we have to look at why Simpson was in Hawai'i to begin with. He was a Governor General of Hudson's Bay Company and he had gone to Honolulu to scout out business opportunities. When he was there, he met with several people including Judge Kekūanaō'a to talk about the financial aspects of taxes and import duties. What he learned was that there was a great deal of uncertainty about the Kingdom's sovereign status that had been challenged by many gunboat commanders, and that uncertainty had adversely affected business prospects. Western countries were imposing what they called extraterritorial treaties on non-western countries that impinged on the local ruler's sovereign rights.

Following his talks about the business environment, he went off to Maui to meet with the King. There, he talked with the King and Richard Charlton, the British Consul, who was very much disliked by the Hawaiian chiefs and King. Kekāuluohi sat in on some of those conversations, and it transpired that they had initially asked George Simpson simply to carry letters to the Queen on their behalf, as they knew he was on his way to London. From there, he expanded the idea and suggested that they send their own diplomats, recommending William Richards as the appropriate person. He helped them put together the documents that Richards would carry with him, and also gave them a line of credit on Hudson's Bay Company of £10,000 to fund the journey. He also pledged that, when they arrived in London, he would join Richards in conversation with British authorities to discuss having their sovereign status guaranteed by Britain. Now, from a man that was essentially a stranger to them, that was a great deal of unsolicited help. You can see where Kekāuluohi's gratitude would have come from, and why she chose to give him what was likely one of her most valuable possessions. This context tells you how desperate the situation was and how grateful they all were when Simpson ultimately wrote the King a letter in which he conveyed our belief that this would be our final struggle for existence.[1]

SEB: Susan brings up a great point about this quilt being such a valuable object. How would the ʻahuʻula have been seen in Hawaiian culture, and how did Kekāuluohi perhaps hope that it was received as a gift?

UH: In this constitutional era, the ʻahuʻula takes on a different significance that it had previously, when they were only for warriors, not the aliʻi. The length of an ʻahuʻula symbolised the power of the aliʻi warrior that wore it. Depending on the height of the person wearing it and how long it was, it could take hundreds to thousands of tiny little feathers to create these ʻahuʻula. They primarily come from five different birds; in this specific ʻahuʻula from Kekāuluohi, they are all from honeycreepers. Honeycreepers are very small birds, so they would only pluck a few feathers at a time. The other main birds that they used were the mamo and the ʻōʻō, which are also very tiny birds. They both had yellow feathers, which were rarer than red feathers. On many feather capes, you primarily see red instead of yellow designs, but the exception is Kamehameha's cloak — an all-yellow garment meant to show his power through both its colour and its floor-length size. Only the really, really powerful aliʻi who ruled over significant areas could marshal the kinds of resources to create an ʻahuʻula like that. Oftentimes these ʻahuʻula were added onto over generations. It becomes an accumulation of the power in that particular lineage.

Not just anybody earned an ʻahuʻula, you had to be of significant status and show your prowess in war. ʻAhuʻula were also worn into battle. In fact, the ones that had been damaged through battle were seen as more significant — as

1   "Instructions from Kamehameha III to Richards and Simpson," 08 April 1842, Ladd Arbitration (Honolulu: Government Press, 1846), 42.

having more mana (power) — because it showed that the person who wore it into battle was hurt and yet they survived, they prevailed. The Bishop Museum in Honolulu has blood-stained 'ahu'ula with holes in them from spear points. From some perspectives that might have been a damaged good, but from the Hawaiian perspective that gave it more mana. One of the ones that Kamehameha gave to King George also had spear point holes and blood on it. These were incredibly powerful gifts and very much political symbols in Hawaiian society. Now, by the time you get to the 1830s, there are no more battles of hand-to-hand combat — instead, the battle was to keep Hawaiian sovereignty, to maintain our autonomy and attain equal status among the world's nations. As an object of political significance, then, we can only speculate that Kekāuluohi saw Sir George Simpson as a warrior in that particular battle.

Pitt Rivers Museum has three 'ahu'ula. 'Ahu is anything that can be worn over you, and 'ula refers to the red colour. So, as I said earlier, the primary bird used for 'ahu'ula feathers was the mamo and the second one was the 'ō'ō, where the yellow feathers came from. There are a handful of other birds that you can get red feathers from — the 'i'iwi, the 'apapane, and 'amakihi for instance — and that's why the red was a little bit easier to obtain than the yellow. Why feathers? Feathers were used in anything that you wanted to invoke akua (deities). In the Hawaiian worldview, height is significant, as anything that can go up into the heavens is considered of godly status. Even though akua translates to God in your standard English dictionary, the word God is actually too limiting. Akua are environmental phenomena; they are essentially anything that cannot be controlled by humans.

Different types of birds are used for different kinds of products. Like I said, those five birds were for 'ahu'ula specifically, as it was harder to marshal those resources. Anything that was harder to get was reserved for the ali'i, the ones with the most power.

SEB: The story about this 'ahu'ula is not just a story about Hawai'i or Kuhina Nui in Hawai'i. It's really a global story. This object has travelled halfway across the world with a mo'olelo, with a story, of its own. ▨

Left: Hawaiian 'Amakihi (male)
Original photo by Bettina Arrigoni

Shirley E. Buchanan is a lecturer at the University of Hawai'i at Mānoa.
J. Susan Corley is an independent scholar.
Catherine 'Īmaikalani Ulep is a PhD candidate in History at the University of Minnesota.
Uluwehi Hopkins is a lecturer at the University of Hawai'i–West O'ahu.

Vanishing Voices

Melissa Price

In this interview excerpt from Dr Melissa Price, Melissa tells us about efforts to preserve Hawai'i's endangered native bird species and the challenges they face.

Most of our native birds at lower elevations have died out due to avian malaria originating from the introduction of mosquitoes, which are not native to the Hawaiian Islands. This is exacerbated by a population of invasive birds that are tolerant to avian malaria, so then spread the disease.

The combination of introduced birds and introduced mosquitoes kept that disease circulating through the population. Most of our students may have seen native water birds or native seabirds, but all of the forest birds they experience in urban areas are all non-native. We really want them to be able to recognise and appreciate our native birds, as it's very hard to have a desire to save things that you don't know or that you're not familiar with.

In Hawaiian culture there is the concept of pilina (relationship). In many indigenous cultures, humans and nature are not viewed as separate — they're viewed as intertwined and reciprocal. Humans take care of nature and nature takes care of humans.

Left: 'I'iwi (Scarlet Honeycreeper)
Original photo by Gregory "Slobirdr" Smith

You may have experienced that when you have an owl that nests in your backyard, its sound brings you calm and peaceful feelings. Or maybe you have an experience with an animal that shows up after a loved one passes away, or a particular plant in your yard that you love cultivating. We build a relationship with our natural world.

For many that live in urban areas, they've lost these relationships with the natural world. They don't follow the moon cycles, which are so influential on our moods and hormone cycles. We want to reconnect people with the birds that are native to their home, that they may not have had the opportunity to personally experience yet.

In February 2022, folks with the Kaua'i Forest Bird Recovery Project got a permit from the State to collect the last five 'akikiki from the wild. They spent about a week searching and only found one juvenile. It's not to say that there aren't others that they might still be able to find, but they're collecting every last one and moving them into captive breeding. Part of the reason for this was that they had recorded as many as two hundred mosquitoes at Kaua'i. With mosquitoes reaching the highest elevations due to climate change, the Kaua'i Forest Bird Recovery Project knew that they had very limited time.

To be able to make meaningful change, we need to have the sort of optimism that brings people together. The ideal responses for these crises might have been possible fifty years ago, but today we're left with few perfect solutions. For example, whilst we have been able to house the 'akikiki in captive rearing, what we've since found is that the males have performance anxiety. If they can hear each other, they won't mate. As such, we're in need of large facilities that would allow the males to sing and produce offspring without being intimidated by the other males.

We've got a long road ahead of us in figuring out a successful captive rearing programme and getting the money for the facilities. Additionally, after the captive breeding, we don't really have anywhere safe to put them. There's another species, the 'akeke'e, on the same island that is at severe risk of extinction. They nest in very tiny rifts in the cliffs, but there is no equivalent habitat within the Hawaiian Islands. They need to do a mating dance that goes up one hundred feet in the air to do a fancy twirl and all of this, and it's really difficult to make breeding facilities for that.

There are some species that have very specific requirements. They can't just evolve their way out of this, especially with how rapidly climate change is taking place. With some of those species we might be able to get creative, but for others, we may not have solutions. We'll have to watch them go extinct, which is heartbreaking. 🔲

Above: Oreomyza bairdi ('Akikiki)
Aves Hawaiienses: The Birds of the Sandwich Islands c. 1890–99
Drawn, lithographed and hand-coloured bird plate by F.W. Frohawk

Dr Melissa Price is an Assistant Professor in the Department of Natural Resources and Environmental Management at the University of Hawai'i at Mānoa.

Kula

Fields
& Plains

"Kalo (taro) is an important food for Hawaiian peoples. In Hawaiian mythology, it is said that kalo is the older brother of humans."
— Kimi Kumagai

The quilts *Kalo* (top), *Hua ʻAi* (middle), and *ʻUlu* (bottom) all represent staple crops and fruits in Hawaiʻi. All of these foods can be found in the kula region of the ahupuaʻa. The kula are the flat and sloping lands between the mountains and the sea that are the main region for farming. The kula can be thought of as the bread basket of the ahupuaʻa. It is also the primary region of the ahupuaʻa where people reside and that houses are built.

A notable feature of the kula is the loʻi kalo, which are irrigated terraces for growing kalo or taro, a root crop. Kalo is a staple in the Hawaiian diet and is pounded to make poi. It is especially dear to Hawaiians because, within local understanding, kalo is considered the elder brother to humans. Hawaiian moʻolelo tells of Wākea, Father Heaven, who bore a child with the Daughter of Earth. The infant, Hāloa, was born prematurely in the shape of a bulb. Wākea buried Hāloa in a corner of his house. The couple's second child, also named Hāloa, was born healthy and would later become an ancestor of the Hawaiian people. Hāloa was to respect and look after his older brother, the elder Hāloa, for all of eternity. The elder Hāloa was known as the root of life and would always sustain and nourish his younger brother and his descendants. At the core of this story is the sentiment of aloha ʻāina (love of the land).

However, colonial interests and commercialisation of the land disrupted aloha ʻāina, particularly the flow of wai (water), leading to drastic changes to the 'breadbasket' and food security. In Hawaiian, the word for wealth is waiwai — the life blood of the ahupuaʻa system. Read on to learn about the fight at Kaʻala Farm to return water to their kula and revitalise a community that was deprived of their waiwai. ▨

Under the Shadow of Mount Ka'ala

Eric Enos

Eric Enos is the Executive Director of Ka'ala Farm Inc. Here, he dives into the Native Hawaiian journey of restoring O'ahu taro farms and reviving an essential cultural practice.

Aloha and welcome to Ka'ala Farm, a cultural learning centre in Wai'anae Valley. My name is Eric Enos. I'm the Co-Founder and Director. We've been up here since the early 1970s. This land was set aside during the Model Cities programme, which was part of Lyndon B. Johnson's great war on poverty. It allowed targeted communities to put resources into social, economic and environmental causes along the coast.

Our community of Wai'anae and the wider Wai'anae moku (island district) has the lowest-performing schools, highest unemployment, the highest per capita number of people in the prison system and a severe lack of educational resources. This qualified us for federal dollars, and so we were part of a youth programme back in the early 1970s that worked with disadvantaged youth on this land. It was once a ranch, but became state lands. The Hawaiian history of the Wai'anae moku goes all the way from Kahe to Ka'ena. The Wai'anae coast is about 18–20 miles in length and our fishing grounds are some of the best on O'ahu because of our deep harbour — the Ka'ena Point — where there are upwelling currents. Our oceans are pretty good, but are now subject to a lot of overfishing.

Like in so many shared native histories, the worst lands
(i.e. the dryest) were the lands set aside for us. Although
this area was historically the "breadbasket" of O'ahu, the
diversion of water for the sugar plantations in the late
19th–20th century and the urbanisation of native people
led to the loss of our traditional lands. Our water rights
were taken away. This was done in the turmoil of the
overthrow of the Hawaiian Kingdom. Then came the sugar
plantations, and our population decreased. It's a similar
story with other Indigenous peoples — without an effective
resistance to diseases, we were dying by the thousands.
The drying up of the land also meant that all the native
people ended up in the outskirts of Honolulu; many of us
soon lost our connection to the land and the language.

I'm from this area. I went to a private school for
native Hawaiians, which was a very interesting school in
that all the teachers at that time were white Protestant. It
was a good, well-meaning, Western education, but taught
us very little about our cultural roots. I never knew Hawaiian
history growing up — the overthrow of Queen Lili'uokalani,
nothing. It was only when I went to college that I really
questioned: "Hawai'i was overthrown? You mean it was an
independent kingdom?" This history was either glossed
over or not taught at all. Thankfully, this has since changed,
but the fact remains that for a native Hawaiian community,
we had very little background in our own culture. I was later
part of the Hawaiian Renaissance movement in the 1970s.
There was a social and environmental resistance to the
bombing exercises by the US Army at Kaho'olawe, which
aligned with the anti-war movement taking place at that
time, most notably against the Vietnam War. It was a kind of
awakening that was happening across all Indigenous people.

When I first came up here as a youth worker, these
ninety-seven acres of land were made available to us. Totally
dry, not one drop of water. We were determined to get back
to the land and our plants — the traditional way of growing

things. We needed to learn our culture of the land, the ʻāina. We started hiking up the mountains of Kaʻala, following the old hunting trails and going over rock walls. If you go up high in our mountain, it's wet. When the rains come, they fall at Mount Kaʻala — the highest mountain on Oʻahu — right above us. We walked up to where the plantations diverted the water and saw the old diversion ditches. After the plantation left Hawaiʻi to find other places for cheaper labour, the Board of Water Supply took over the system and piped the water out.

We started doing more research and I saw all these rock walls, so I called some friends at the Bishop Museum. They came and showed me a map that was done in 1906 by MD Monsarrat, a surveyor who was active in Hawaiʻi for around thirty years. It showed the entire back of the valley in taro cultivation and, well, taro needs water. Therefore, we deduced that all of these dry lands were once rich with water. This 1906 map also showed all the diversion ditches and plantations. I walked to the source of the water, where the Board of Water Supply was taking it out, and I thought, what gives them the right? How do we get our fair share of this water, to be returned back to our community? This question goes into Hawaiʻi's Water Law. Our aliʻi were very well-educated, and although Queen Liliʻuokalani was overthrown, our traditional Indigenous practices were embedded in the Constitution.

A friend of ours introduced us to Senator Richard Matsuura, the Head of the Governor's Agriculture Coordinating Committee. He came to visit and we went to the plantation diversion ditch where they were taking the water out of the mountain, and brought it all the way down to the sugar fields below us. He pointed at a ditch where a little bit of water was leaking out to form a stream, and told us, "you're going to get that water." He explained that we had to do a Conservation District Use Application (CDUA) which meant agencies, research and hydrologists. We needed to have everybody on board with this application. He set us up with a wastewater engineer at the University of Hawaiʻi, and they helped us design this system. The hydrologist explained that the stream was surface runoff

Above, left-to-right: Thea Correia, Stone Perez,
Eric Enos and Marenka Thompson-Odlum at Ka'ala Farm
Photo by Marenka Thompson-Odlum

water, which happens because the Board of Water Supply can't capture all of the water — some escaped into the stream. It didn't go all the way down the mountain, but it went down far enough. The engineer designed a plan to re-route that water using a two-inch PVC pipe to take it one mile down the mountain to the farm. I also worked with the forestry department, and I had the archaeologists come in to make sure we were not going to disturb anything. We were just hoping to lay our pipe on top of the land, making sure not to affect anybody downriver of us. I submitted the CDUA, but a couple of months later, it was denied.

In 1978, there was a dispute between Waiāhole taro farmers and the Board of Water Supply. The Board had gone up into one of the valleys and capped the stream and were taking the water out. Taro farmers on the windward side were still growing taro, but their stream was reduced. The Board of Water Supply responded that it was their job to provide water for everybody, but that their mandate

Above: Kaʻala Farm
Photo by Marenka Thompson-Odlum

was mainly to meet domestic and urban dwellers' water needs. The Waiāhole taro farmers took the state to the Supreme Court. William Richardson, chief justice of the Hawaiian State Supreme Court at that time, ruled in favour of the farmers because it was in the Constitution. It states that water is held in trust — it's not a commodity, you can't own water. Our aliʻi were very wise indeed. The taro farmers won the case, a landmark decision by the Supreme Court that declared water is to be held in public trust.

So, that was all happening when we applied for the CDUA in Waiʻanae. Understandably, they were weary of another one of these water rights cases. Even though we were denied the water rights in 1978, we received a donation of pipes. In an act of civil disobedience, we went up there and ran the water down anyway, but we knew nothing about planting taro. Fortunately, during the water dispute we had joined the Waiāhole taro farmers in their protests and supported them. Returning the solidarity, they showed us how to plant taro. This solidified the bond between us and the Waiāhole people.

Two months later, state enforcement agents from the Department of Land and Natural Resources (DLNR) accused of us illegal diversion of water in Waiʻanae Valley. There were a lot of developments that they wanted to happen — hotels, golf courses — but at the cost of destroying our fishing grounds. They were promising jobs, but when the jobs were finished, our reef, our ocean and our fish would be gone. We made a lot of enemies in our own community, those who were promised the construction jobs. The community was divided. When the DLNR accused us of illegal diversion of water, I showed the CDUA and told them what we were trying to do: just taking a little bit of water and putting it back into taro, whilst taking the kids off the streets and planting traditional plants for food and medicine. I was invited to give testimony in front of the Land Board, and they decided to allow us to lay our pipe on state-leased land. We evoked the Waiāhole water decision and the constitutional question of "who owns the water?" It's like asking who owns the sun, or the wind. Thanks to our ancestors and what they had codified in the Constitution, water was for all. It was our right as Hawaiian citizens.

So here we are. That's our water story, and here we are today at Kaʻala Farm. We're still fighting. We're now dealing with invasive species and invasive people. These lands are very difficult to work, but our struggle is who we are. The fact that we can still struggle means that we can also stand up and fight.

Eric Enos is the Executive Director of Kaʻala Farm Inc. Born on the Waiʻanae coast of Oʻahu, amongst the largest population of Native Hawaiians, he comes from a family of builders, farmers, craft folk, caregivers and educators. His life's work is to restore the land and water with his community.

Makai

The Ocean

"I love the beaches and ocean. My favourite quilt is the one I made for my husband, where John designed all the different fish that can be found in the Hawaiian waters. My spirit and my love of the ocean and its dolphins lives in this quilt."
— Mie Tashiro

In the ahupuaʻa system, the ocean is not viewed as separate from the mountains or fields. Its health and vitality is connected to the uplands. The quilts *Honu* ("Turtle"), *Nā Iʻa o Ke Kai* ("Fish of the Ocean"), and *Naiʻa* ("Dolphins") depict various types of ocean life. The border around each of the four samplers of the *Nā Iʻa o Ke Kai* quilt call to mind a unique aspect of the makai in Hawaiʻi: fishponds.

Very few fishponds currently survive in Hawaiʻi. However, the following interviews with members of the non-profit organisation Paepae o Heʻeia tell the story of revitalising an eight-hundred-year-old fishpond in the ahupuaʻa of Heʻeia in Oʻahu. Their work tells the tale of the importance of fishponds, combating food insecurity and the continued effect of plantation ditches and water diversion, as laid out by Eric Enos in the kula region [pp. 84–91]. Learn about the beauty and power of pōhaku (rocks) and about those who call themselves the "weavers of rocks," as they rebuild the fishpond. ▨

Where the Fishpond Breathes

Angela Hiʻilei Kawelo

Angela Hiʻilei Kawelo is a native of Kahaluʻu, Oʻahu and became Paepae o Heʻeia's Executive Director in 2007. Here, she tells us about being raised in Kahaluʻu by her fishing family and being a student of the art and science of lawaiʻa (fishing) her entire life.

There were once more than four hundred fishponds throughout Hawaiʻi. Oʻahu alone was home to just shy of a hundred fishponds; places like Puʻuloa (Pearl Harbour) and Waikiki counted around fourty, while Kāneʻohe Bay, right here in Heʻeia, had just shy of thirty. They are now under threat.

Our mission at Paepae o Heʻeia — a non-profit that cares for and stewards the Heʻeia fishpond — is to implement the values and concepts from the model of a traditional fishpond to provide physical, intellectual and spiritual sustenance for our community. In layman's terms, we have been trying to restore this fishpond for the last twenty years. We are aiming to preserve its cultural aspect and the physical structure itself, which is eight hundred years old, but also to restore the community. The purpose of the fishpond is also to nurture fish, to grow seafood and to feed the community.

We see fishponds as representative of the resources of the area, namely water. You couldn't have a fishpond in an estuary if there wasn't fresh water present. The presence and proliferation of fishponds in Hawai'i reflects the amount and abundance of fresh water that was once here. I wish we could say that there was just as much fresh water today, but that's one of our big challenges.

Fishponds were built to cultivate seafood and fish as that was the primary protein for Hawaiian people. We ate a lot of fish. The largest fishponds were in excess of about five hundred acres, however the smallest could be of less than an acre and still provide enough for one 'ohana (family). The pond at He'eia, which at eighty-eight acres is somewhere in the middle of that spectrum, fed the entire local community. Typically, fishponds were built to cultivate herbivores. The keystone fishpond species were 'ama'ama (Hawaiian mullet) and awa (milkfish). That's not to say that it was a monoculture system. There are all kinds of other species in the pond as well — predators, crustaceans, molluscs.

Within the ahupua'a system, fresh water makes its way down from the mountains to the ocean. These shallow coastal systems are where phytoplankton abound, which is the food source for the 'ama'ama and awa. So long as fresh water is making its way down to the ocean, the system feeds itself. However, much of our fresh water has been diverted, taken over the mountain to feed the sugar cane fields in the Ewa plains on the other side of the island. Even though sugar is no more, the systems that were constructed to service sugar are still being used today. We have in the back of He'eia some straw-like wells and tunnels that drill down into our aquifer and move water from this ahupua'a to the neighbouring ahupua'a. We love our neighbours Kaneohe and Kailua, but that's just what it's like here in Hawai'i. The same can be said for other nearby ahupua'a. The water just keeps getting moved; roughly half of the water that should be in our stream is not. Much of the water is diverted for residential use and for watering facilities like golf courses. We should give water to our neighbours and to our people, but there are still things that we can do to tighten it up, like reusing and recycling water.

Within Heʻeia, there are other organisations doing similar work to us. Mauka, upstream of us, are our good friends and our partner organisation. The big picture is restoring not just one fishpond, but an entire ahupuaʻa. Another big challenge is the invasive species that take up a lot of the available water. A lot of the work that we do in the ahupuaʻa is removing invasive, introduced species and replacing them with native species. Organisations upland from Heʻeia, such as Kākoʻo ʻŌiwi, are restoring the taro fields. In a traditional system, fresh water moves through the loʻi system and taro farmers, and then goes back into the stream. Today, it's a little more detached. You have residential communities where streams are channelised and big four hundred acre wetlands that are overgrown with invasive species. The once interconnected system has become segmented and fragmented.

We can often hear the sound of the flow of water through our mākāhā (a sluice gate or weir). We have seven of them here at the fishpond, and they are the lifeline of the pond. It's a structure that moves water — so that both saltwater and freshwater can come into the fishpond — and it's also the mechanism by which we stock the pond with fish. A lot of Hawaiian words you can't really break apart linguistically, but you can with mākāhā: mā is short for māka, or your eye; kā is a current; and hā is breath. Maka also identifies a place where something really important happens. So, mākāhā is the place or the site at which the fishpond is able to breathe. That's the word. 🔲

Right: Heʻeia Fishpond
Photo by Marenka Thompson-Odlum

————
Angela Hiʻilei Kawelo is a native of Kahaluʻu, Oʻahu and became Paepae o Heʻeia's Executive Director in 2007. Raised in Kahaluʻu by her fishing family, she has been a student of the art and science of lawaiʻa (fishing) and Kāneʻohe Bay her entire life.

Weaving Rocks

Keahi Piʻiohiʻa

Keahi Piʻiohiʻa was born and raised in Kailua, and was first introduced to the fishpond in 2011 with UH-Mānoa's Malama Loko Iʻa class before joining the Kū Hou Kuapā at Paepae o Heʻeia in June 2012. In this interview, Keahi Piʻiohiʻa reflects on the art of Hawaiian dry-stacking.

My name is Keahi Piʻiohiʻa. I'm the Restoration Coordinator at Paepae o Heʻeā. Our crew is called Kū Hou Kuapā, which literally means "to let the wall rise again." That's what we do. We're trying to restore an eight-hundred-year-old fishpond to the best of our ability and with what we have. Right now, we're dealing with all sorts of invasive species in and out of the pond, including mangrove and seaweeds. Ultimately, our hope is that the fishpond will sustain Hawaiʻi again.

The moʻolelo we want to share is that there's nothing in Hawaiʻi built in the last hundred years that has lasted as long as the fishpond, but there are fewer surviving fishponds here to tell this story. That's all we really want to do: share the story and get people to believe that there can be thousands and thousands of fish right here again.

Overleaf and right: Heʻeia Fishpond
Photos by Marenka Thompson-Odlum

We live on a beautiful island that's capable of growing food, capable of providing fish, and yet for 90% of our goods we rely on the rest of the world. I have kids and I'm worried for them if they keep losing places like this. We call rocks "pōhaku" in Hawaiian. Po is the darkness; po is where we come from and where we return to. When you're born, you're born through po, and when you pass away, you return to that. Haku is a word that means "creator" or "master." The creation, the haku, and the po, from the dark, refers to all of our islands and volcanoes. The volcano came from the bottom of the ocean and it created a rock that grew and grew until it turned into our islands. For our hōkūleʻa (voyagers) who navigated the oceans, the first time they spot an island from fifty miles offshore it just looks like a rock, but that rock becomes Hawaiʻi.

There's a beautiful song written by Eleanor Prendergast called "Kaulana Nā Pua." She was a lady-in-waiting for Liliʻuokalani. She wrote a verse in the song:

> Aʻole mākou aʻe minamina
> I ka puʻukālā a ke aupuni
> Ua lawa mākou i ka pōhaku
> I ka ʻai kamahaʻo o ka ʻāina.

It states, simply and beautifully, "we are not satisfied with your mountains of money. We are satisfied with rocks, which flourishes our land." That's how important rocks and pōhaku are to our kūpuna. In the form of imu (underground oven), rocks cook our food; as pōhaku kuʻi ʻai (stone food pounder), they prepare our staple foods. Rocks built the walls for our taro, for our kalo. Rocks built fishponds. Rocks built the platforms for our houses. That is something that endures. Everything else may come and go, but these rocks remain. There are rocks out there that were brought here eight hundred years ago, and they're still there. Along with those rocks, we say ʻaumakua, our spirit ancestors, they all still reside here too.

Rock wall-stacking is probably the second major job we do after clearing the invasive mangrove. We reinforce it with rock and coral, a practice which is called the art of stacking walls, or Hawaiian dry-stack. It's called uhau humu pōhaku, which literally means, "the sewing together of rocks." That's how awesome our kūpuna were, to be as poetic as that. When you're doing it, you realise that it's not just a bricklaying type of stack. It's all dry-stack, no mortar; we literally try to sew together the rocks. This is the pōhaku version of a quilt — the interlocking and the interjoining of rocks to make something.

Those rocks are going to noho (to live) for another eight hundred years. Eight hundred years from now, they're going to say, "Eh, these guys came in eight hundred years ago and restored this thing. And we've never looked back since." That's the goal — we are trying to build community more than we're trying to build the fishpond up.

Overleaf: Kilauea volcano, Hawai'i Island
Photo by Marenka Thompson-Odlum

Keahi Pi'iohi'a is the Restoration Coordinator at Paepae o He'eia. Born and raised in Kailua, Keahi was first introduced to the fishpond in 2011 with the University of Hawai'i at Mānoa's Malama Loko I'a class before joining the Kū Hou Kuapā at Paepae o He'eia in June 2012.

Conclusion

Marenka Thompson-Odlum

The *Nā Mele ʻo Hula Kahiko* quilt (left) depicts the instruments that traditionally accompany hula dancers — the pahu hula (drum), ʻulīʻulī (feathered rattles) and the pūniu (coconut knee drum). In contrast, hula ʻauana (contemporary hula) is known for its Western influences, including the use of musical instruments such as the guitar and ukulele. The word ʻauana literally means "to wander" or "to drift," used here to communicate a drift away from the more sacred elements of hula kahiko (ancient hula). ʻAuana has become what most people globally think of as hula, solidifying the practice in perceptions of Hawaiian culture.

    The journey of *Mauka to Makai: Hawaiian Quilts and the Ecology of the Islands* has highlighted the shifting sands of culture and of "authenticity." Hawaiian quilting, like hula ʻauana, is a 19th–20th century creation that developed in response to outside influences — ever-changing but rooted in the knowledge of the kūpuna (ancestors). Notions of 'authenticity' and indigeneity are often problematically entangled.

When I first began this project, I faced many queries about how authentic the quilting was to Hawaiian culture. My choice of working with a practice that has a relatively short history (two hundred years) in the Hawaiian Islands was similarly called into question. I was reassured by a conversation I had with Hawaiian artist Solomon Enos, the son of Ka'ala Farm's Eric Enos. As an artist whose work leans towards the graphic and the futurist, he too often faces the question of authenticity. His answer is simple: "I am continuing the work of my ancestors, just in a different medium." That is the beauty of this culture. It can remain rooted in worlds of knowledge, whilst also making space for more people, ideas and means of expression — weaving the threads between the old and the new, the past and the present and the yet-unseen future.

There is a Hawaiian proverb, "I ka wā ma mua i ka wā ma hope" ("The future is in the past"), which describes the way in which many of the answers and solutions for our current questions reside in long-held traditions and knowledge. The saying hones in on a Hawaiian concept that runs throughout the stories represented in this book: the interconnectedness of all things, including time.

Many of the stories told in this text came about due to a disruption of that connectivity. Quilting was first introduced to Hawaiians by white American and European missionaries as a way to counteract nudity, which they viewed as a problem. At this time, Hawaiians already made their own fabric by the way of kapa (barkcloth), but this practice has since been disrupted and was for a long time thought to be lost. The ecological and social issues of species extinction, lack of water, wildfires and food insecurity as chronicled by these interviews can also be viewed through the lens of disruption — the disruption of a finely-connected and balanced ahupua'a system. The physical, ecological and ideological colonisation of the Hawaiian Islands was perpetuated through the disruption of long-held knowledge.

Amongst all these weakened threads, the beauty remains that it is difficult to erase that which is so deeply rooted. Aloha 'āina (love of the land, of place) persists. Hawaiian quilting is proof that continued aloha ʻāina amidst monumental disruption can help to usher in a new form, a new quilting style which is specific to the archipelago. Still grounded in traditional Hawaiian aesthetics, the quilts pass down stories of the land, the people, resistance and resurgence.

Like these quilts, Kaʻala Farm and Paepae o Heʻeia are forging new paths in the face of disruption by looking to the past. In Heʻeia, the definition of an ahupuaʻa may have slightly expanded, but the spirit of community and mālama (responsibility) — both to your human and non-human neighbours — remains at its core.

The journey behind the creation of these fifteen quilts is centred on these threads of connectivity, weaving new ones, pulling on existing threads and imagining their possible pathways. I liken my Hawaiian quilting journey to the echo lines of a quilt: it began with one object, and over the last five years has reverberated to produce this book and create a beautifully diverse network of people and knowledge.

Sol Enos once told me, "the role of the hula dancer is as important as the community's weavers and kapa makers, for the fabric of reality also needs frequent upkeep." It reminds me that everyone and everything has a purpose. The Poakalani quilting group, like the hula dancer, is stitching together the moʻolelo of a place, of a people, of an ʻohana.

A Hawaiian Glossary

'A'ohe pau ka 'ike i ka halau ho'okahi
Proverb meaning "all knowledge is not taught in the same school" or "one learns from many sources."

Āholehole
Hawaiian flagtail fish.

'Ahu'ula
Feather cloaks worn in battle and also used as diplomatic gifts. ('Ahu refers to anything that can be worn, and 'ula means red.)

Ahupua'a
An ancient land division system divided into strips of land from the mountain to the sea, supporting self-contained communities.

'Āina
Land, earth;
Hawaiian ancestral land.

'Akikiki
Hawaiian honeycreeper endemic to Kaua'i.

'Akua
Deities, Gods.

Ali'i
Chiefs, nobles, Hawaiian royal family.

Ali'i Wahine
Female chiefs.

Aloha 'Āina
Love of the land.

'Ama'ama
Mullet, indigenous fish.

'Amakihi
A common species of Hawaiian honeycreeper.

'Anae
Mullet, indigenous fish.

'Apapane
A small, deep crimson species of Hawaiian honeycreeper.

'Aumakua
Spirit ancestors.

Awa
Milkfish.

'Awapuhi 'Ula 'Ula
Red Ginger plant.

Hālau Kuiki O Owyhee
Quilt (kuiki) class (hālau) of Hawai'i (Owyhee).

Hāloa
Child of Wākea (the father of Heaven) and Papahānaumoku (the mother of Earth), born in the shape of a bulb.

He'e
Octopus.

Heiau
Hawaiian temple or shrine.

Hi'iaka
Goddess of hula and younger sister of the goddess Pele.

Ho'iho'i
To return or restore.

Ho'iho'i ka wai
The return of water.

Hōkūle'a
Voyagers. Also Arcturus, the navigational star (hōkū meaning "star," and le'a meaning "gladness.")

Honu
Turtle.

Hua'ai
Fruit.

Hula
The dance of Hawai'i.

Huli
Taro shoot, as used for planting.

Humuhumunukunukuāpua'a
The state fish of Hawai'i, a species of triggerfish.

'I'iwi
Scarlet Hawaiian honeycreeper.

'Ike
Knowledge, or to greet or recognise.

Imu
An underground oven.

Kahawai
A stream or river.

Kāhili
Feather standard, symbolic of Hawaiian royalty.

Kai
Sea.

Kākū
Great barracuda.

Kalo
Taro, a staple Hawaiian root vegetable.

Kānaka Maoli
Native Hawaiian flag; a full-blooded Hawaiian person.

Kānāwai
Law.

Kapa
Barkcloth fabric, made from wauke (paper mulberry) or mamaki bark.

Kapa kuiki
Hawaiian quilting.

Kauila
An endemic species of tree in the buckthorn family.

Kekāuluohi
Kuhina Nui (second-in-command) of the Kingdom of Hawai'i from 1839–45. Also the wife of both Kamehameha I and II.

Kia'i
Guardian or protector.

Kinolau
The many forms taken by a supernatural body; a physical manifestation of Hawaiian deities.

Koa
Warrior; also a native tree whose wood is used to build canoes and other items.

Kōkala
Porcupinefish.

Konohiki
Headman of an ahupua'a (land division) under its chief.

Kū Hou Kuapā
"To let the wall rise again."

Kuapā
Rock wall of a fishpond.

Kuhina Nui
A unique and powerful position within the Kingdom of Hawai'i between 1819–64, second-in-command to the King.

Kukui 'o Hale Ali'i
"Lights of the House of Royalty," referring to 'Iolani Palace, Honolulu.

Kula
Fields, plains.

Kumu
Teacher.

Kūpuna
Ancestors; elders, grandparents or older people.

Laka
The goddess of hula.

Lei
Flowers, leaves, shells or feathers that are strung and/or entwined and given as a symbol of affection.

Liholiho
Birth name of King Kamehameha II.

Liliko'i
Passion fruit.

Limu
Edible algae, seaweed.

Lo'i
Wetland, or irrigated terrace.

Lo'i kalo
Irrigate terraces for growing kalo (taro).

Loko i'a
Hawaiian fishponds.

Loko i'a kuapā
Walled coastal fishponds.

Mahiole
Hawaiian feather helmets.

Maka'āinana
People of the land.

Maka'āinana wāhine
Women of the land.

Mākāhā
Sluice gate or weir. (Mā – short for "māka" or eye; Kā – a current; Hā – breath.)

Makai
Towards the sea.

Mamo
Black Hawaiian honeycreeper.

Mana
Spiritual or supernatural power.

Manakō
Mango.

Manu
Bird, or any winged creature.

Mauka
Towards the mountain or uplands.

Mea no'eau
Skillfully created works.

Mele
Song or anthem.

Moi
Threadfish.

Moku
A district of an island; a small offshore island; or a poetic reference to large islands.

Mo'olelo
Story, tale, myth, history.

Muliwai
Estuaries.

Nā I'a o Ke Kai
Fish of the ocean, or ocean life.

Nā Koa
Hawaiian warriors, or "the courageous."

Nā Mea Ali'i Wahine
That which belongs to queens or chieftesses.

Nai'a
Dolphins.

Noho
　To live, reside, inhabit.

Oʻahu
　The most populous of the Hawaiian
　Islands, and the seat of Honolulu.

ʻOhana
　Family, kin, group.

ʻŌhelo
　A small native shrub in
　the cranberry family.

ʻŌhiʻa ʻAi
　Mountain apple.

ʻŌhiʻa Lehua
　Flower of the ʻōhiʻa tree.

ʻŌlelo
　Language or speech.

Oli
　The act of chanting; also, a chant
　that is not danced to, especially
　with prolonged phrases chanted
　in one breath, often with a trill
　at the end of each phrase.

Olona
　A native shrub, the bark of
　which was a source of strong,
　durable fibre for fishing nets.

ʻŌʻō
　A black honeyeater.

ʻŌpae
　Shrimp.

Palani
　A surgeonfish, renowned
　for its strong odour.

Palila
　An endangered grey, yellow and
　white Hawaiian honeycreeper.

Pāpaʻi
　A crab

Pāpio
　A fish.

Peʻahi
　To wave or fan.

Pele
　Lava flow, named for the
　volcano goddess Pele.

Pilina
　Relationship or connectedness.

Pōhaku
　Rock or stone.

Pōhaku kuʻi ʻai
　Stone food pounder.

Puaʻa
　A pig.

Pualu
　A species of surgeonfish.

Puhi
　Eels.

Pūniu
　Coconut knee drum used in
　classical hula dancing.

Ua Mau ke Ea o ka ʻĀina i ka Pono
　"The life of the land is perpetuated
　in righteousness," or "our
　sovereignty and balance has
　been restored." Phrase famously
　uttered by King Kamehameha III
　when Hawaiʻi's sovereignty was
　reinstated from the British in 1843.

Uhau humu pōhaku
　"The sewing together of
　rocks," referring to the
　art of stacking walls.

Uka
　Inland, upland, mountains.

ʻUlīʻulī
　Feathered rattles used in
　classical hula dancing.

ʻUlu
　Breadfruit.

Wai
　Water.

Waiʻanae
　"Waters of the mullet,"
　a district of Oʻahu.

Waiwai
　Wealth, goods, property.

Wākea
　Father Heaven, the mythical
　ancestor of all Hawaiians.

Wao akua
　The realm of the Gods. A distant
　mountain region, believed to
　be inhabited only by spirits.

Wao kanaka
　The realm of man. An inland region
　where people may live, usually
　considered below the wao akua.

Wāwae Moa
　Chicken feet; a traditional
　stitch of Hawaiian quilting.

The Poakalani Quilters

This book is published with heartfelt gratitude to the Poakalani quilting family. We thank the Poakalani 'ohana for their exquisite contributions to the cultural tapestry of quilting. Your knowledge and artistry not only enriches the collection of Pitt Rivers Museum, but also fosters connections across oceans and generations. For more information on Hawaiian quilting styles, techniques, classes and more, please visit www.poakalani.net

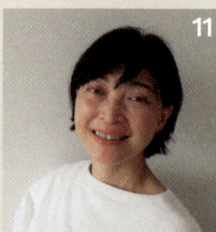

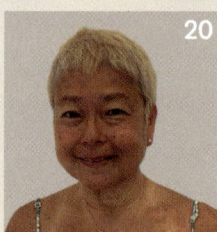

1	Midori Andrews	11	Nobuko Nakagawa
2	Chikako Asano	12	Hana Yoko Nakayama
3	Rae Correia	13	Kathi Nakayama
4	Eriko Furukawa	14	Anne-Marie Naughton
5	Pat Gorelangton	15	Yuko Nishiwaki
6	Takako Jenkins	16	Cissy Serrao
7	Tomoko Kato	17	Wakako Shionoya
8	Kimi Kumagai	18	Susie Sugi
9	Susan Lessa	19	Yoshimi Suzuki
10	Jennifer McCullough	20	Mie Tashiro

21   Pat Gorelangton working on the *Tennessee Iris* quilt as part of the *50+1 State Flowers* quilt project. Saturday Poakalani quilting class at the Higashi Hongwanji Mission, Honolulu. August 2023. Courtesy of Marenka Thompson-Odlum.

22   John Serrao and Tomoko Kato outside the Hawai'i State Archives, Honolulu. Courtesy of Poakalani & Co.

Marenka Thompson-Odlum is Research Curator (Critical Perspectives) at Pitt Rivers Museum, University of Oxford. She grew up in the Caribbean nation of St Lucia. At Pitt Rivers Museum, Marenka leads an Art Fund project to commission new objects for the Museum's collections, build new relationships with indigenous communities, and enhance the Museum's displays. This project has led her to work with artists and makers from the islands of Hawaiʻi, Haida Gwaii and Hokkaido.

Marenka is also the lead researcher on the *Labelling Matters* project, which investigates the problematic use of language within Pitt Rivers Museum's displays and re-imagining the definition of a label with a decolonial lens. Her doctoral research at the University of Glasgow explored Glasgow's role in the trans-Atlantic slave trade through the material culture housed at Glasgow Museums.

Marenka has, notoriously within the Poakalani quilting circle, been working on her very own 22x22 inch Hawaiian quilt for the last three years, with the end still nowhere in sight. Part of her curatorial practice is to understand the collections she helps care for by learning the methods through which they were made. This has led her to pound kapa (barkcloth) in the sunshine of Waimānalo, plant taro in shadow of Mount Kaʻala and birdwatching in the ʻōhiʻa lehua forest near the crater of Kīlauea. She hopes that this book will inspire others to learn about and continue this beautiful tradition.

Common Threads Press is a small press that specialises in the radical histories of crafts and making.

Established in 2019, we publish zines and curate events that uplift marginalised histories of creative work. Our publications are written collaboratively with early-career researchers, writers, students and academics from all around the world who share our love for craft histories.

More from Common Threads Press:

Stitching Freedom:
Embroidery & Incarceration
Isabella Rosner
978-1-91632-347-6

Rights Not Charity:
Protest Textiles & Disability Activism
Gill Crawshaw
978-1-91632-346-9

Slow Grown:
Plants, Folklore & Natural Dyeing
Ciara Callaghan
978-1-39992-032-2

Diasporic Threads:
Black Women, Fibre & Textiles
Sharbreon Plummer
978-1-39991-944-9

Many Hands Make a Quilt:
Short Histories of Radical Quilting
Jess Bailey
978-1-91632-348-3

Copyright

All artworks © Pitt Rivers Museum, University of Oxford.

p. 17   2022.57.13 Kukui 'o Hale Ali'i
p. 19   2022.57.3 Nā Koa
p. 21   2022.57.15 Nā Mea Ali'i Wahine
p. 23   2022.57.1 'Ōhi'a Lehua
p. 25   2022.57.7 Kalo
p. 27   2022.57.14 'Ulu Poi
p. 29   2022.57.2 Nā Mele 'o Hula Kahiko
p. 31   2022.57.12 Ti Leaf and Laua'e
p. 33   2022.57.11 Pe'ahi
p. 35   2022.57.4 'Awapuhi 'Ula 'Ula
p. 37   2022.57.5 Bird of Paradise
p. 39   2022.57.6 Honu
p. 41   2022.57.8 Dolphins
p. 43   2022.57.10 Nā I'a o Ke Kai
p. 49   2022.57.9 Hua 'Ai

Photographs

Unless otherwise stated, all photographs courtesy of Marenka Thompson-Odlum.

The publishers have made every effort to trace the copyright holders of the works illustrated and apologise for any omissions or errors that may have been made.

Text

The interview, "An Introduction to Hawaiian Quilting and the Poakalani 'Ohana," (pp. 8–13) was first published in *Spaces of Care: Confronting Colonial Afterlives in European Ethnographic Museums* (edited by Wayne Modest and Claudia Augustat, 2023, Transcript Verlag.)

Published by Common Threads Press
www.commonthreadspress.co.uk

ISBN: 978-1-06862-500-8

Senior Editor: Laura Moseley
Editorial Assistants: Marisa Clements & Eleanor Gaffney
With thanks to Vanshika Poddar

Design: Chris Shortt

Typefaces: Greed (Martin Vácha)
Gulax (Morgan Gilbert)
CM Geom (Cedric Müller)
VTC Harriet (Vocal Type)

Printed and bound by Short Run Press, Exeter, Devon, UK

MIX
Paper | Supporting responsible forestry
FSC   FSC® C014540
www.fsc.org

© Common Threads Press 2024